Papillio Press
Sacramento, Calif.
2014

The Mystery of Tree Ring Dating
by Steven D Holmes

When I was an undergraduate, I took a botany class. The most memorable thing that I learned was how tree rings could be used to date things. The study of that is, "dendrochronology". It seemed real cool to me that if I had the rings of one tree, I could match them up to the rings of a dead tree, and continue doing that with older and older trees to get a "look" at history. It sounded good at the time and I believed it.

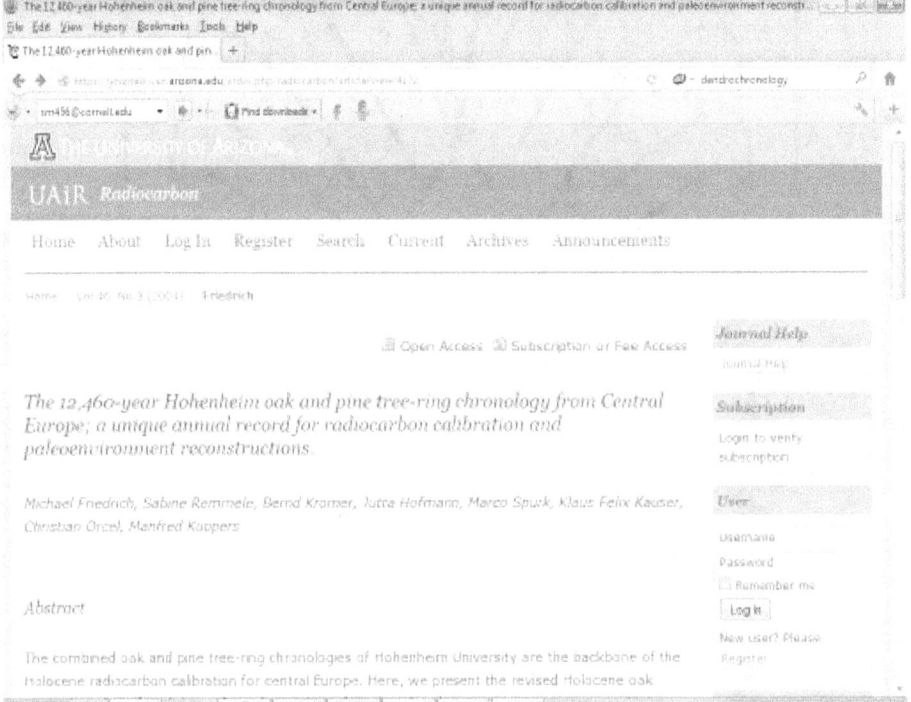

Some decades later, "learned" how tree rings were supposed to be used to look back as far as 12,460 years. It was supposedly done mainly with two trees, matching an overlap of 538 rings from one old tree to one that was much older. It was hard to find but I found this site with a PDF file that gave lots of information:

In this file, I found the chart below. I didn't see any supposed correlation between the top and bottom data.

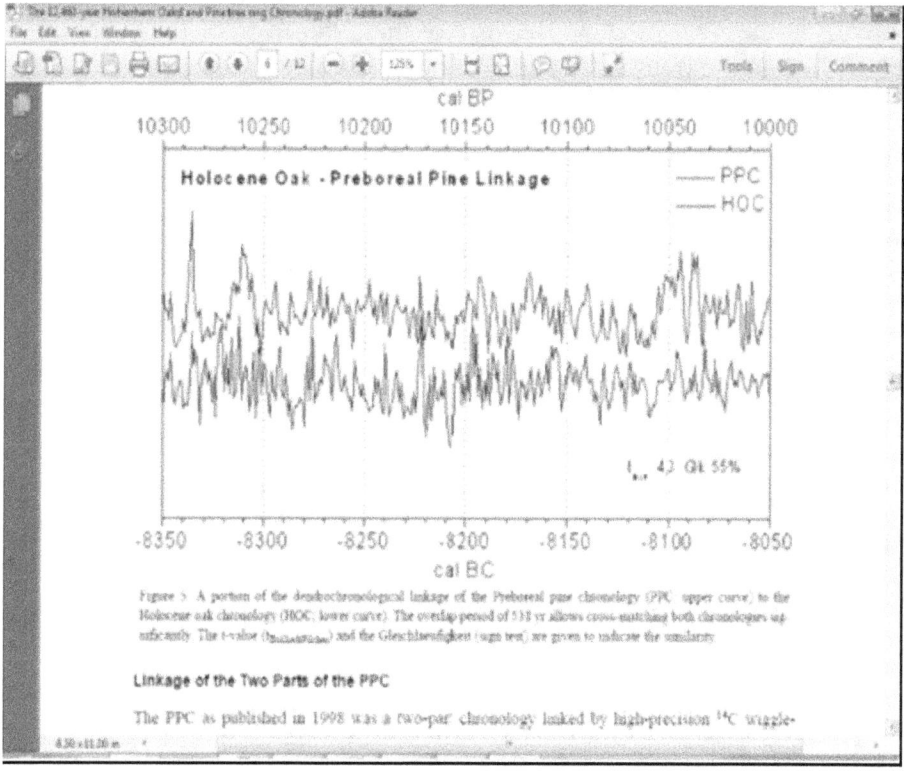

I live in Sacramento and in the local McKinley Park, a couple of elm trees succumbed to Dutch Elm disease. I decided to measure the rings. Since I didn't have polished specimens where all the rings were so easy to count, I followed from one ring to the other, usually ending up with not looking at what made a straight line. I also decided to take photos of the rings for future reference.

We have a place in Lake Tahoe somewhat frequently so I decided to also get data from trees there. In the neighborhood, I measured and took photos of some trees. Then I was very lucky to find in the state park with the Baldwin estate, a tree that was much older and the date of when it fell was also known, 1996.

Sacramento data:

What was really striking to me was how if I just looked at the ring comparison of the trees for the last fifty years, they didn't show the same trends in the rings' widths. Because of that, the rings for the years up to 2012, when the trees were cut, are shown first.

Elm tree number 1 shows narrow rings from around 1992 to 2012

Elm tree number 2 shows the widest rings for the last five years, rings narrowing for about ten years, narrow rings for another ten years, and rings that got wider.

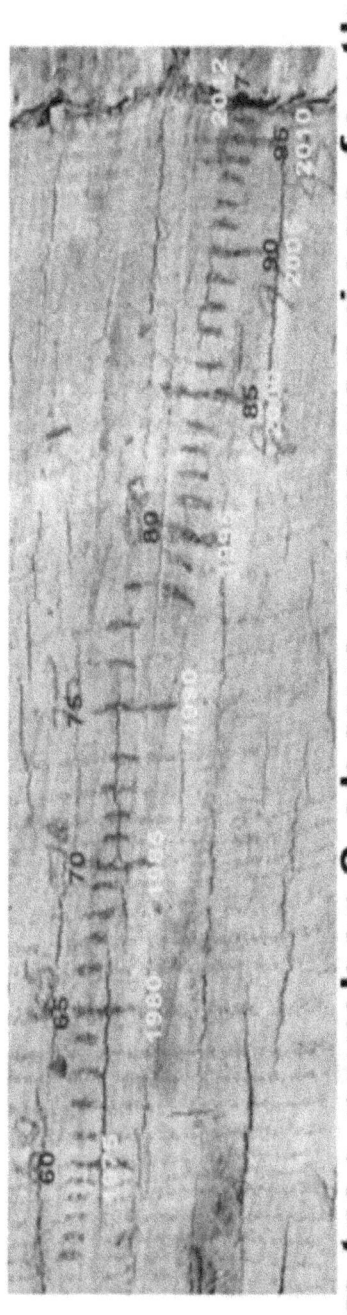

Elm tree number 3 shows narrower rings for the last fifteen years, wider rings for seven years, and then narrower rings again.

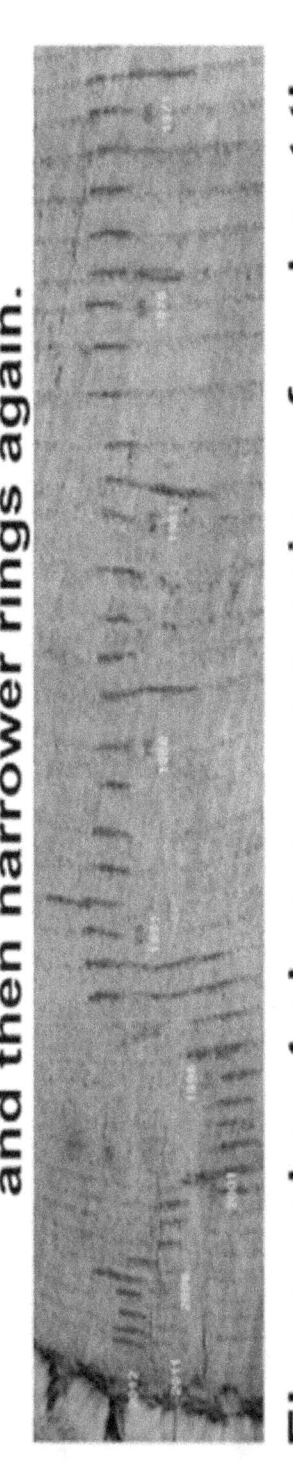

Elm number 4 shows narrow rings for about the last seventeen years and wider rings that get much wider over fifty years ago.

Elm tree number 5 shows narrow rings for the last twelve years and wider rings before that.

Elm tree number 6 shows narrower rings for the last fourteen years, wider for twelve years before that, and much wider rings before that.

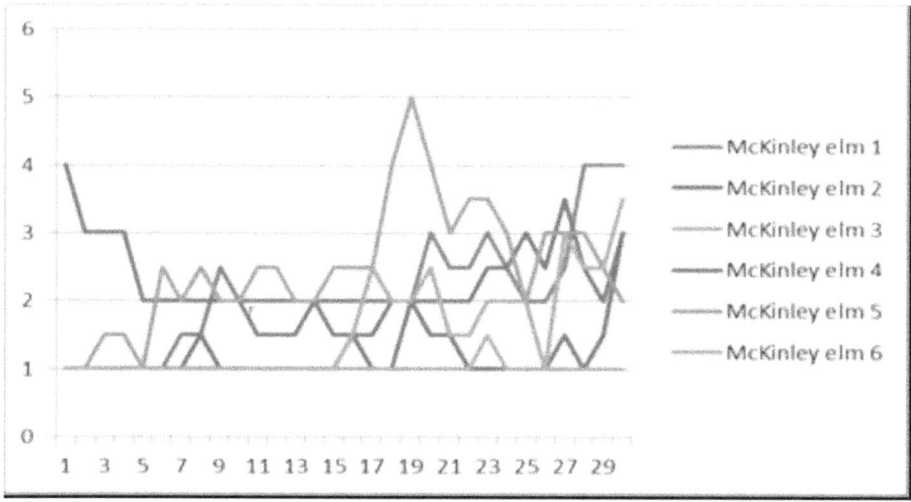

Here is a graph of the data. There doesn't seem to be a correlation. I decided to send this data off to experts. I searched online and found a web site with lots of information about dendrochronology. I sent off this set of photos with the rings marked to see if somebody might be able to explain the lack of correlation. Unfortunately, nobody responded by this time.

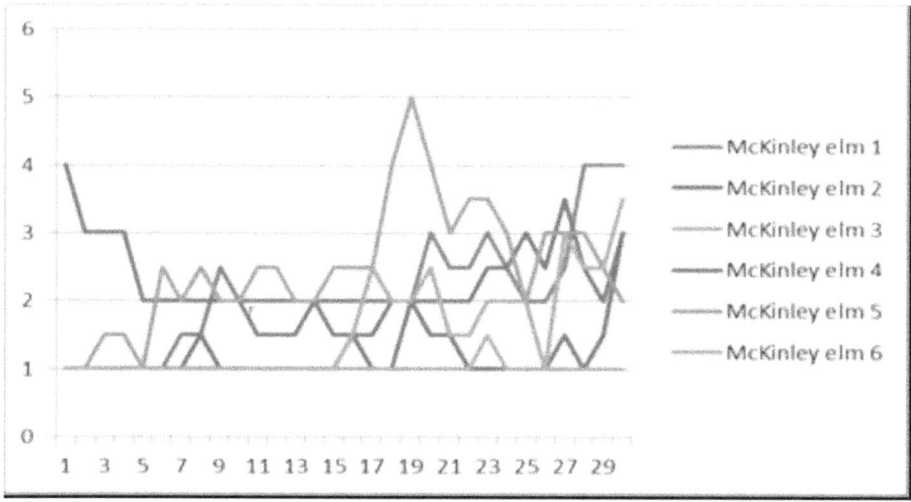

Lake Tahoe data:

Tim Sizemore's cedar tree on Gardner Mt., South Lake Tahoe, cut down in 2012 shows good growth for the last ten years, less for ten years before, and back to good growth.

In Lucky Baldwin's tree cut down in 1996 on the south end of Lake Tahoe, the rings are wider from 1988 to 1993 but pretty equal otherwise.

If we examine Mr. Sizemore's tree (2 pages back) we see that 1995 had a very large growth with half as much on the years before and afterward.

But on Lucky's tree (below) it is hard to see what the rings were like between 1993 and 1996. There is large growth before 1993 and not afterward.

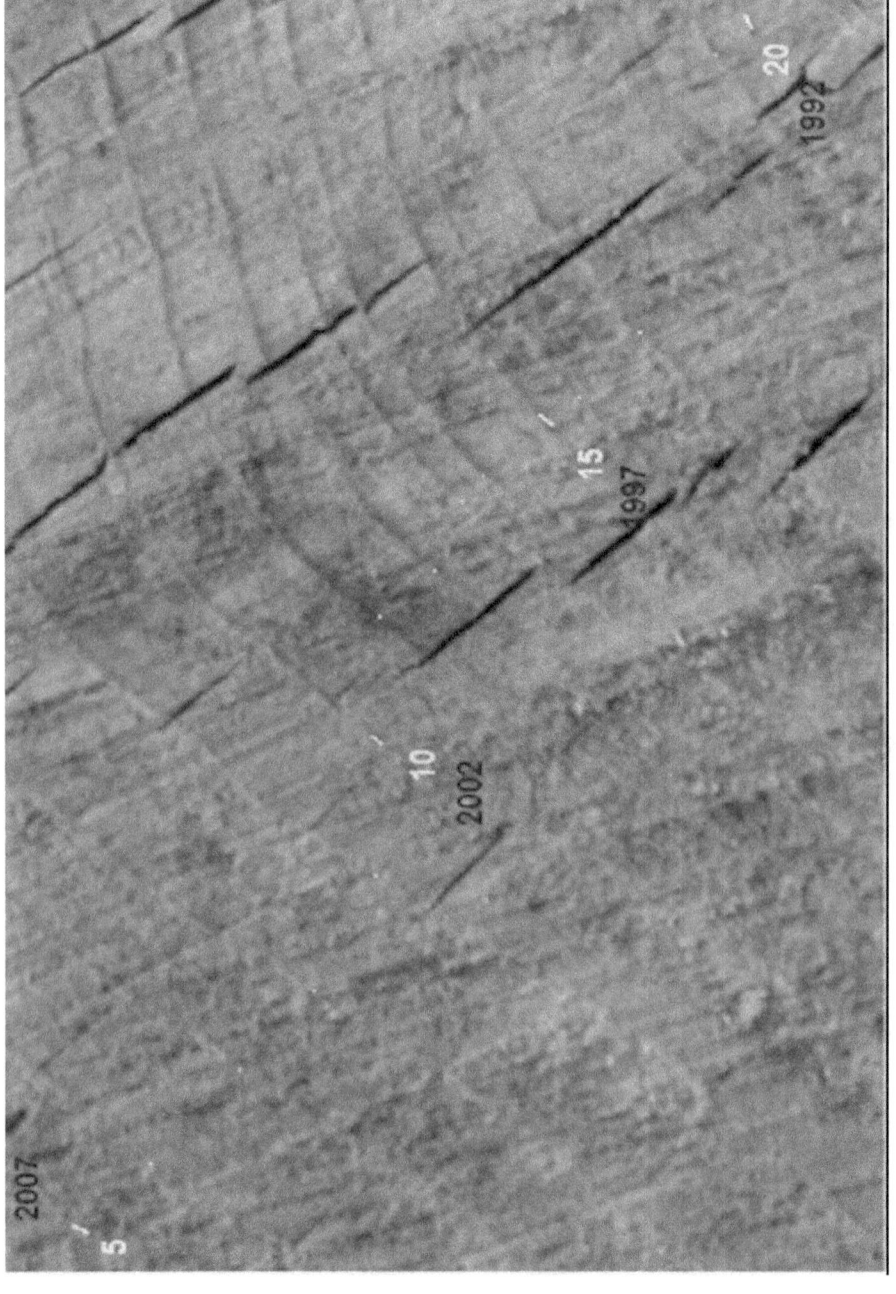

As the saying goes, "proof is in the pudding". If this theory is true, I ought to be able to figure out when a tree was cut down by comparing its rings to another tree. In the Tahoe area, there are lots of tree stumps to use in that way. A neighbor, Todd, cut a tree down not long ago. I don't recall the exact year, but it was six to ten years ago. So, if this method works, it ought to be easy to figure out when the tree was cut down, right?

If we look at the rings on Todd's tree, we can see that the rings got progressively larger for the last thirty-five years and then thirty-six years ago the rings were a lot narrower. Since that tree was cut down between six and ten years ago, that smaller ring would be forty-two to forty-six years ago, which would be 1974 to 1980. On Todd's tree, the narrower rings are 1953-1958.

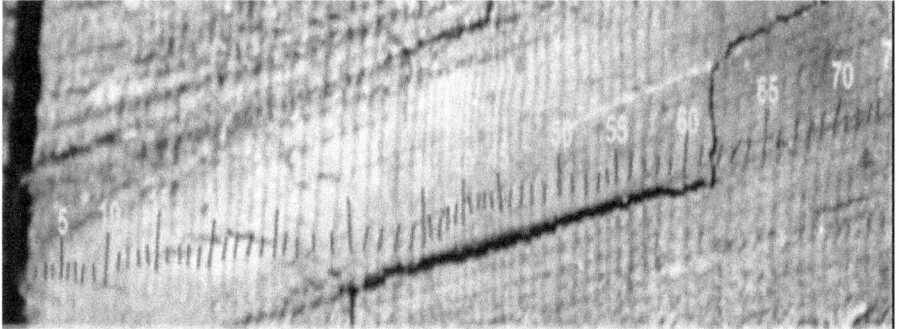

So, let's try another tree (next two pictures). On this one in the neighborhood, the year that the tree was cut down is unknown but since it's in the same residential area of Todd's tree, it was probably cut down within the last three decades (but due to the newness of it, I'm guess it was cut down in the last fifteen years).

When we examine the rings of this tree, the rings are narrower between the last ten years before the tree was cut down. Also, the rings were much wider from thirty-six to fifty-two years before being cut down, a period of sixteen years. So, if tree ring widths are reliable, we ought to be able to find some other nearby tree with at least ten years of narrow rings (years 0-10 before this tree was cut), twenty-six years of moderate rings, then sixteen years of wide rings before then got narrow for about eighteen years, and then pretty wide rings for forty-five years.

That didn't match up with Todd's tree: last 10 years narrow, 25 next are moderate width, 10 of narrow, 29 of wider, 48 of wider and wider rings, one narrow ring 123 years ago, and then 41 years of wider years back to the most early growth.

Lucky's tree rings didn't match the unknown tree or Todd's tree: 3 moderate, 5 wider, 30 moderate, 5 narrower, 94 moderate, and 25 wider.

Tim's tree is even more erratic and doesn't match any of the other trees: 3 narrow, 7 wider, 3 moderate, 1 narrow, 1 moderate, 1 narrow, 1 moderate, 1 wide, 1 moderate, 1 narrow, 3 moderate, 1 wider, 1 narrow, 6 wider, 3 moderate, 1 narrow, 1 wider, 1 narrow, 1 moderate, 1 wide, two moderate, and 2 narrow, which takes it back to 1969.

There is also a large stump near highway 89 and the road to Valhalla. I have counted 414 rings on it. I don't know when it was cut down, but in asking park volunteers it was likely at least a few decades since it was cut down. If dendrochronology is real, then it ought to be useful to figure out the date that this tree that is within a half mile of Lucky's tree was cut down.

Going backwards in time, the tree ring widths (shown above) were: 35 years narrow, 11 years moderate, 12 years wide, 1 year moderate, 1 year narrow, 2 years moderate, 1 year wide, 12 years moderate, 5 years narrow, 7 years very narrow to take us back to 295 years before the tree was cut down.

Following are pictures of other South Lake Tahoe tree rings for comparison. Look to see if there is a pattern of large and small rings similar to the other South Lake Tahoe trees.

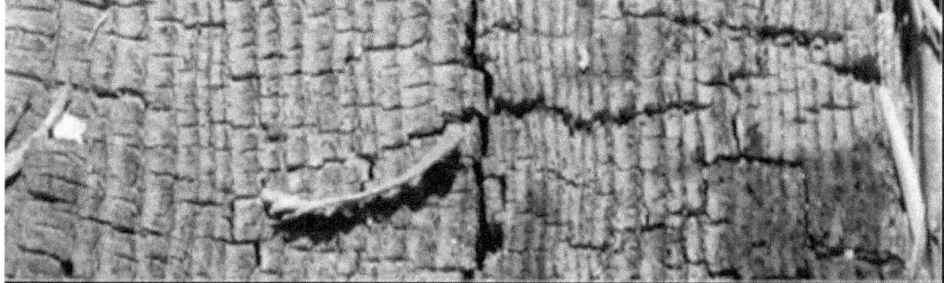

This stump was behind the high school

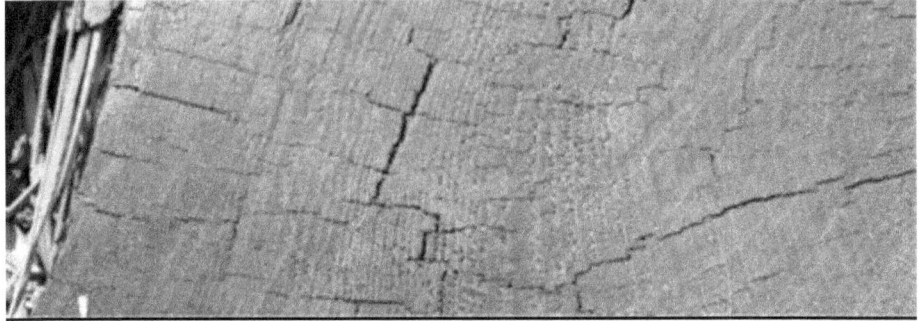

This stump was also behind the high school.

This stump was on the Baldwin estate.

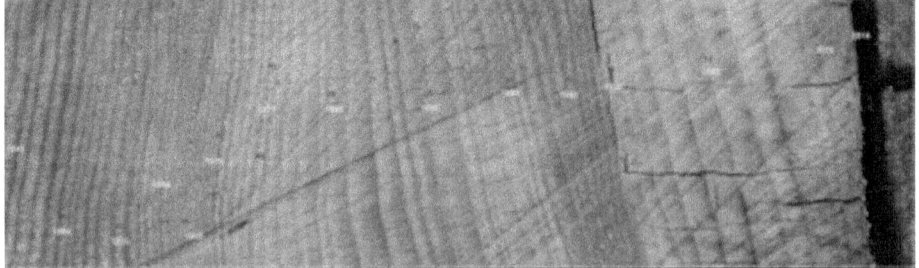

Another Baldwin estate stump.

What is left of a stump near the Baldwin estate, felled so long ago that most of it rotted away.

The remains of an old stump, felled many years ago, along the bike trail near Baldwin Beach.

Another old stump fairly close to the Baldwin estate.

What is left from another stump near the Baldwin estate.

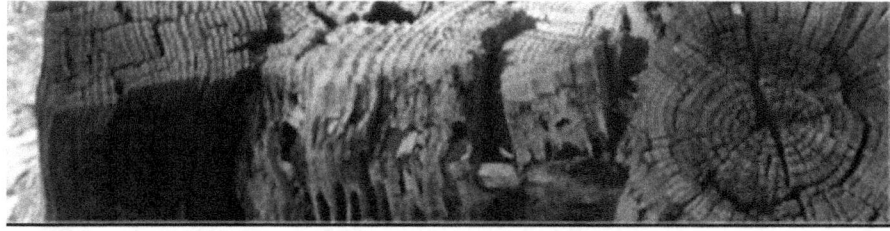

An 87-year-old stump near Baldwin with the last 21 years narrow.

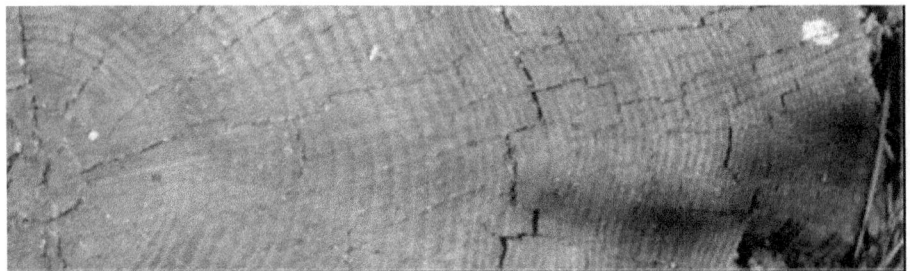
Another stump near the Baldwin estate.

A 103-year-old stump near the Baldwin estate.

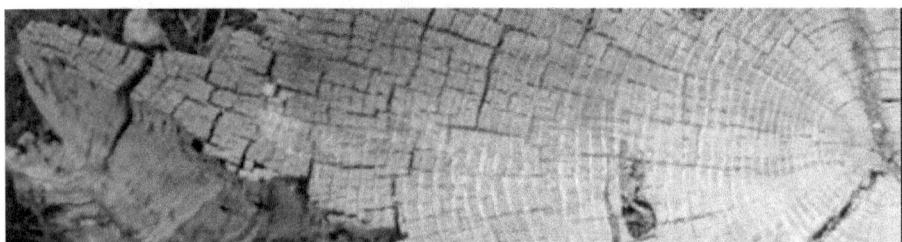
Rings on a 95-year-old stump near the Baldwin estate.

A stump near the Baldwin estate with 128 years of rings countable and the last 76 narrow.

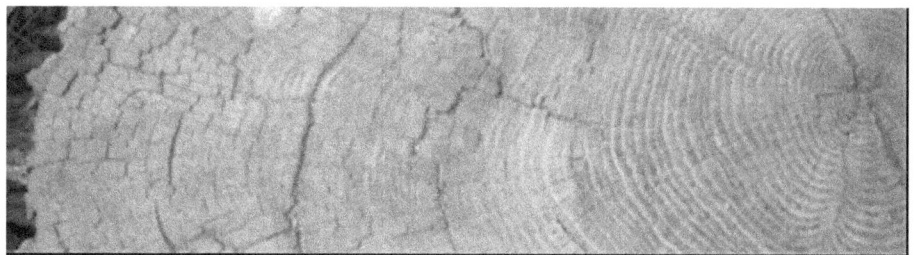

A 121-year-old stump near the Baldwin estate with the last 81 rings narrow.

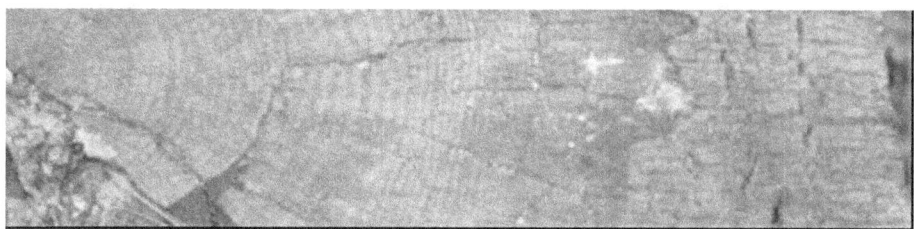

A 152-year-old stump near the Baldwin estate with the last 99 years narrow.

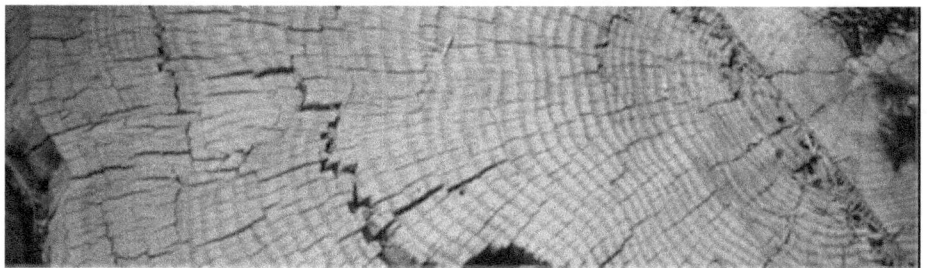

An 88-year-old stump near the Baldwin estate with the last 28 years narrower.

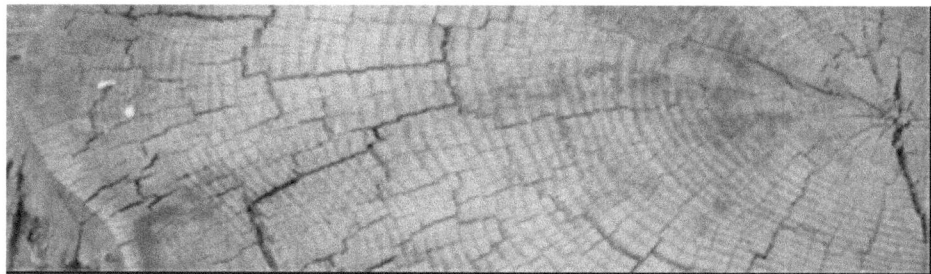

A 105-year-old stump near the Baldwin estate with the last 19 years narrower.

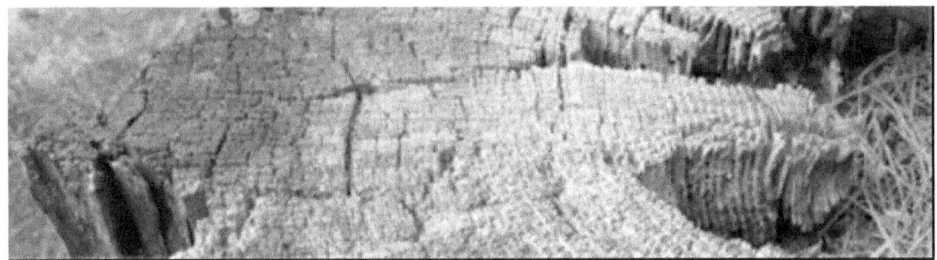

What's left of another old stump near the Baldwin estate.

What's left of another old stump near the Baldwin estate.

What's left of another old stump near the Baldwin estate.

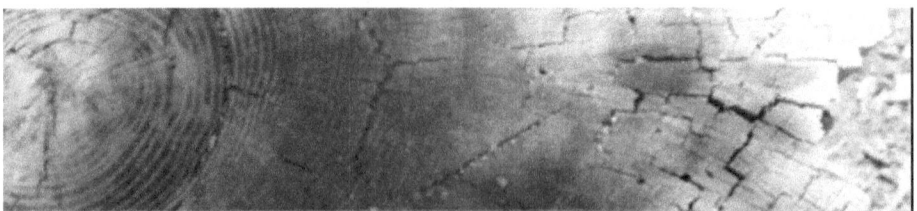

Rings from the last old stump that I found near the Baldwin estate.

This stump was also on the Baldwin estate.

Here are some stumps from trees killed and subsequently cut down in the Angora fire in South Lake Tahoe in 2007. These are all stumps of pretty old age and all within a 200x200 area so they all ought to have similar rings for the last years, right?

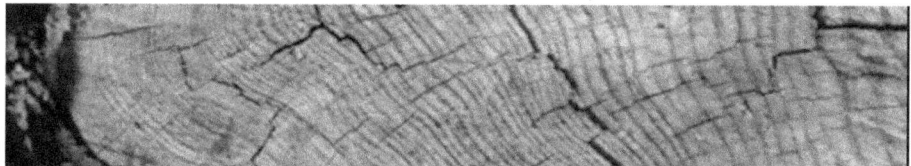

A stump near the water tanks on Gardner Mountain.

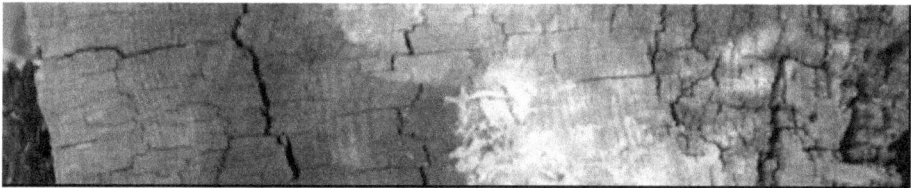

Another stump up near the Gardner Mountain water tanks.

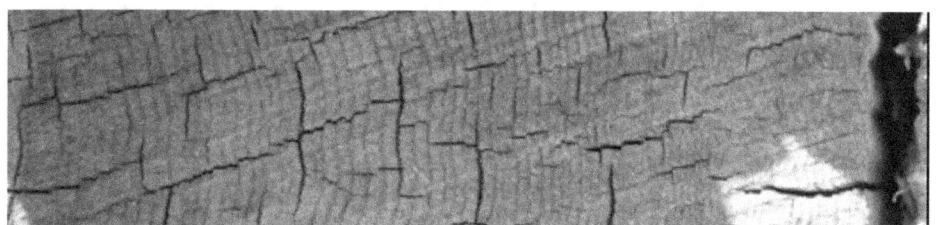

A third stump near the Gardner Mountain water tanks.

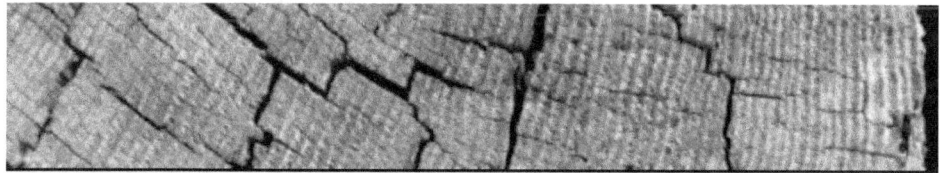

A fourth stump near the Gardner Mountain water tanks.

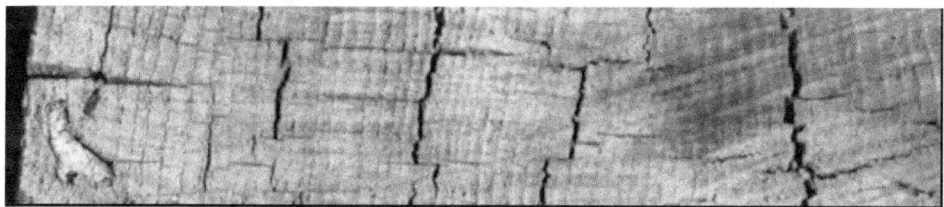

A fifth stump near the Gardner Mountain water tanks.

Along the bike trail in South Tahoe, between 15th Street and the road to Pope Beach there was a reforested area. Due to change in forest policy, they have decided that it's best to reduce the number of trees by something that looks like ninety percent. Therefore, many trees were cut down in 2013. Here are pictures of their rings. They are all within 200 feet of one another, though their ages vary quite a bit. Look to see if there is an correlation between their rings for their last years of living.

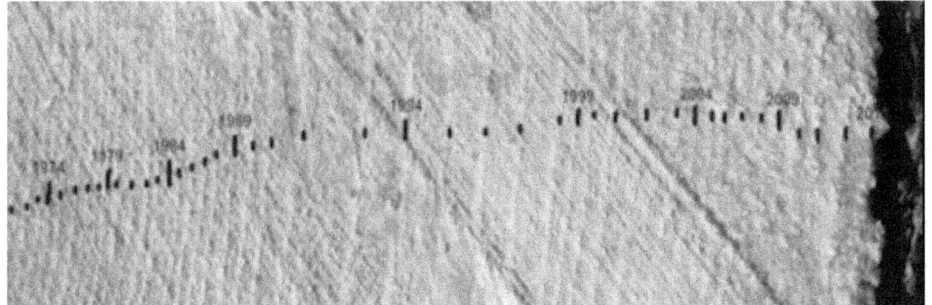

#1 – 2013 to 1979 medium width rings (34 years), 1979 to 1991 wide rings (12 years), 1979 to 1959 narrower rings (20 years), and wider for some older years.

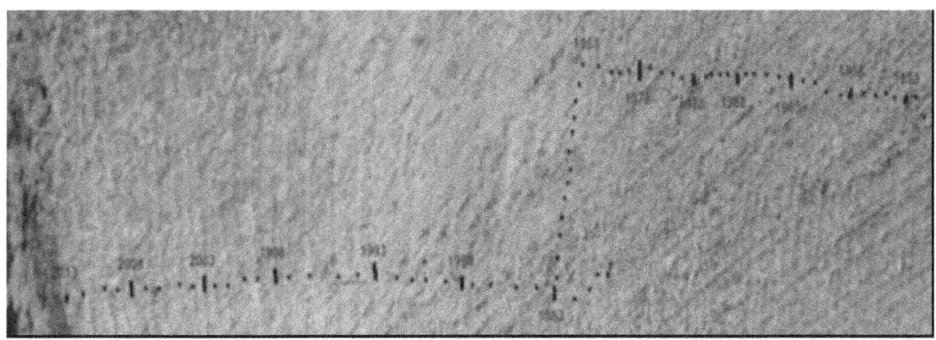

#2. – 2013 to 2011 wide rings (2 years), 2011 to 1997 narrower rings (14 years), 1997 to 1995 wide rings (2 years), 1995 to 1981 narrower rings (14 years), 1981 to 1952 years narrow (29 years), 1952 to 1948 wider rings (4 years), 1848 to 1941 narrow rings (7 years), wider rings before that.

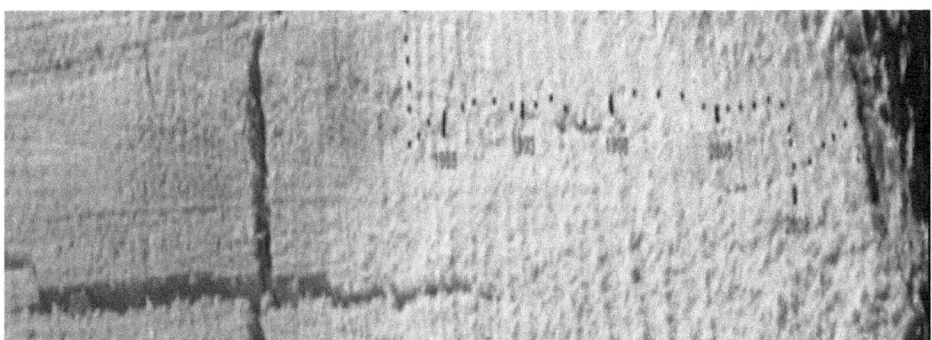

#3 – 2013 to 2002 medium rings (11 years), 2002 to 1996 wider rings (6 years), 1996 to 1986 narrower (10 years), 1986 and earlier narrow rings.

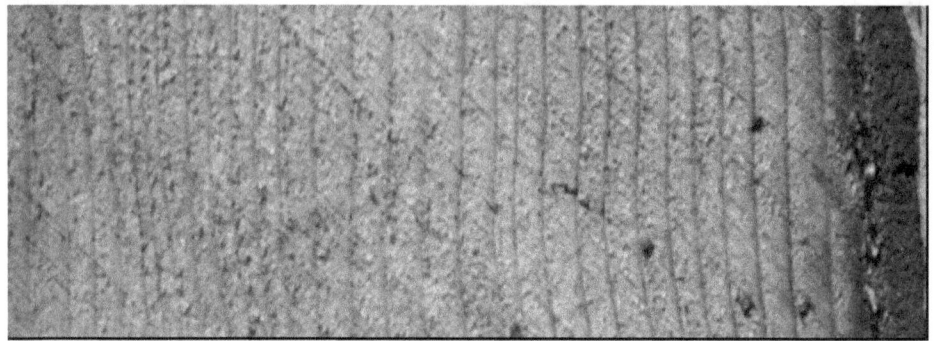

#4 – 2013 to 1995 wide rings (18 years), 1995 to 1888 narrower rings (7 years), 1988 to 1983 wide rings (5 years), 1982 to 1976 medium rings (6 years), 1976 to 1918 years narrow (58 years).

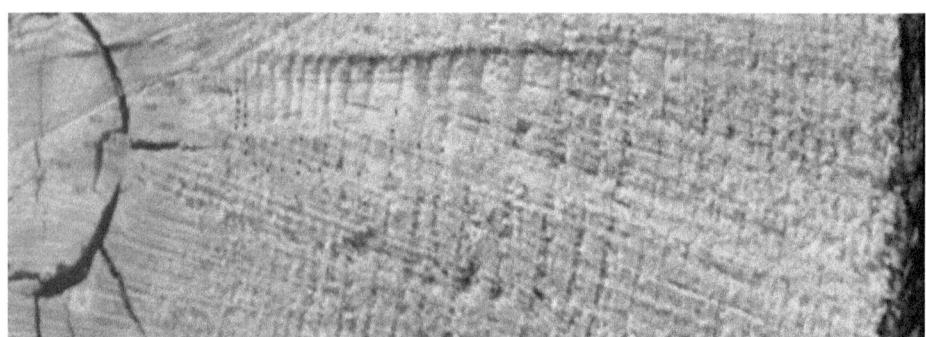

#5 – 2013 to 2011 wide rings (2 years), 2011 to 2001 medium rings (10 years), 2001 to 1995 wide rings (6 years), 1995 to 1990 medium rings (5 years), 1990 to 1982 narrower rings (8 years), 1982 to 1912 very narrow rings (70 years).

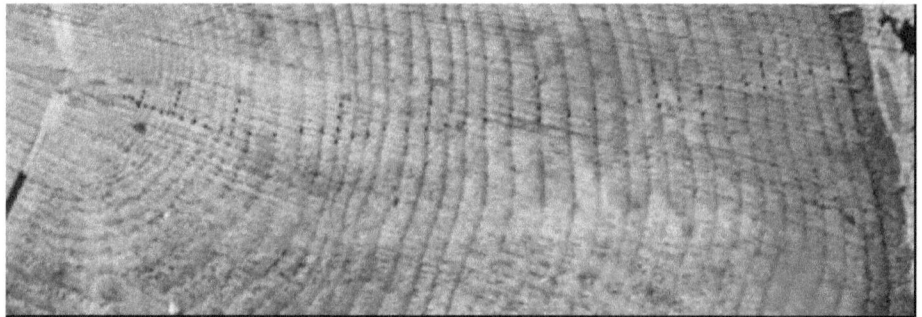

#6 – 2013 to 2001 medium rings (12 years), 2001 to 1996 wide rings (5 years), 1996 to 1987 medium rings (9 years), 1987 to 1984 wider rings (3 years), 1984 to 1978 narrower rings (6 years), 1978 to 1968 even narrower rings (10 years), 1968 to 1947 very narrow rings (21 years).

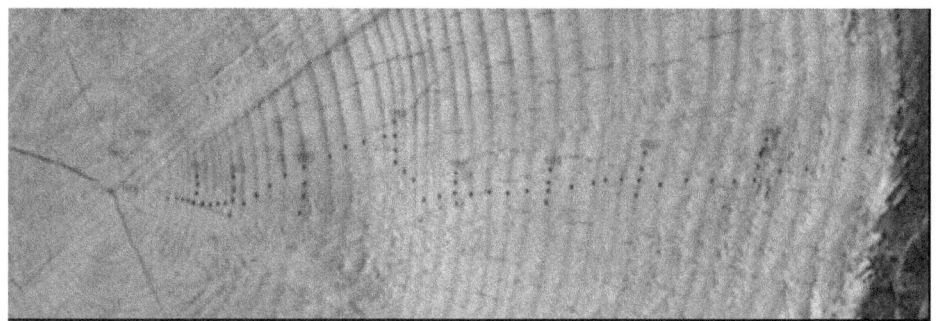

#7 – 2013 to 1994 wide rings (19 years), 1994 to 1992 medium rings (2 years), 1992 to 1988 narrow rings (4 years), 1988 to 1981 medium rings (7 years), 1981 to 1968 narrower rings (13 years), 1968 to 1954 narrow rings (14 years).

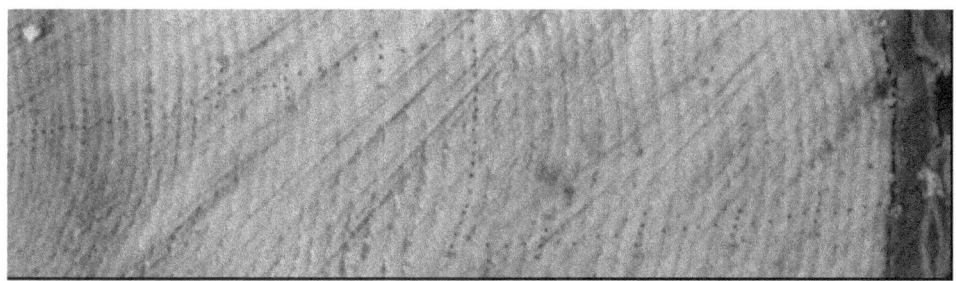

#8 – 2013 to 1988 wide rings (25 years), 1988 to 1987 medium ring (1 year), 1987 to 1982 wide rings (5 years), 1982 to 1961 medium rings (21 years), 1961 to earlier years narrow.

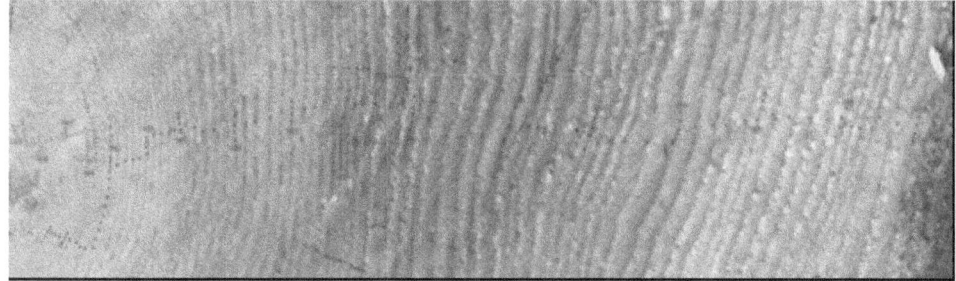

#9 – 2013 to 2012 wide ring (1 year), 2012 to 2011 medium ring (1 year), 2011 to 2007 narrow rings (4 years), 2007 to 2003 medium rings (4 years), 2003 to 2002 narrow ring (1 year), 2002 to 2000 medium rings (2 years), 2000 to 1997 wide rings (3 years), 1997 to 1993 narrow rings (4 years), 1993 to 1991 medium rings (2 years), 1991 to 1990 narrow (1 year), 1991 to 1983 medium rings (8 years), 1983 to 1964 narrower rings (19 years), 1964 to 1952 even narrower rings (12 years), 1952 to 1933 very narrow rings (19 years).

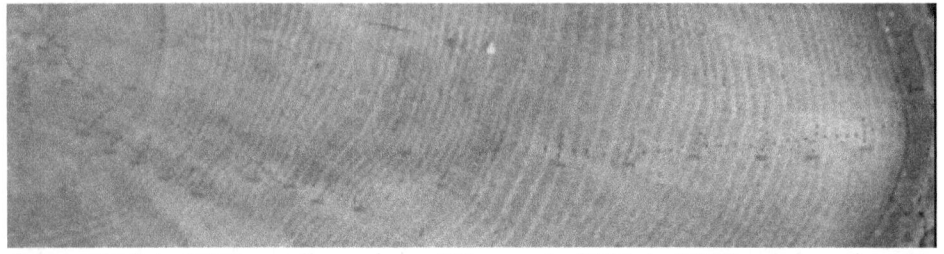

#10 – 2013 to 2001 medium rings (12 years), 2001 to 2000 narrow ring (1 year), 2000 to 1986 wider rings (14 years), 1986 to 1981 wide rings (5 years), 1981 to 1979 medium rings (2 years), 1979 to 1901 narrow rings (78 years)

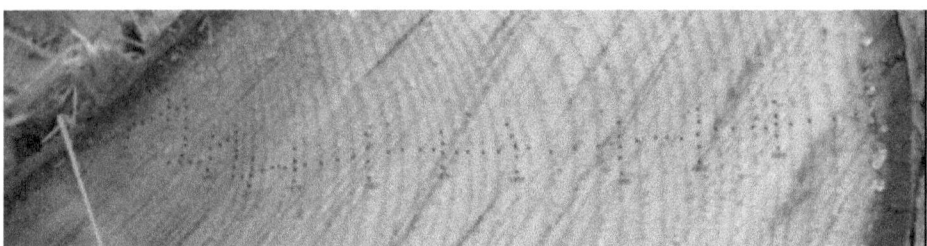

#11 – 2013 to 2009 wide rings (4 years), 2009 to 1997 medium rings (12 years), 1997 to 1994 wide rings (3 years), 1994 to 1988 medium rings (6 years), 1988 to 1986 narrower ring (2 years), 1986 to 1983 little wider rings (3 years), 1983 to 1977 narrower rings (6 years), 1977 to 1933 narrow rings (44 years).

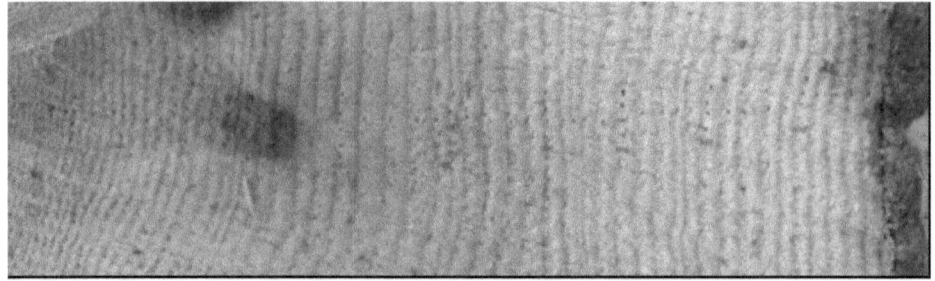

#12 – 2013 to 2002 medium rings (11 years), 2002 to 2001 narrow ring (1 year), 2001 to 1997 medium rings (4 years), 1997 to 1995 narrow ring (2 years), 1995 to 1989 medium rings (6 years), 1989 to 1988 narrow ring (1 year), 1988 to 1985 medium rings (3 years), 1985 to 1981 wide rings (4 years), 1981 to 1977 medium rings (4 years), 1977 to 1969 narrower rings (8 years), 1969 to 1951 narrow rings (18 years).

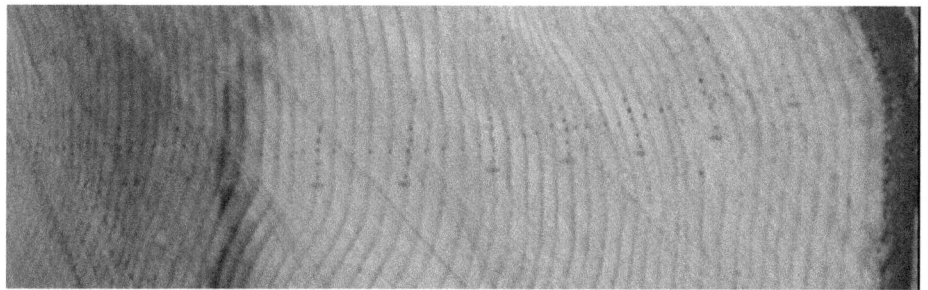

#13 – 2013 to 2011 wide rings (2 years), 2011 to 2002 medium rings (9 years), 2002 to 2000 narrow rings (2 years), 2000 to 1995 medium rings (5 years), 1995 to 1990 narrow rings (5 years), 1990 to 1979 medium rings (11 years), 1979 to 1972 narrower rings (7 years), 1972 to 1958 even narrower rings (14 years), 1958 to 1934 narrow rings (24 years), 1934 to 1900 very narrow rings (34 years).

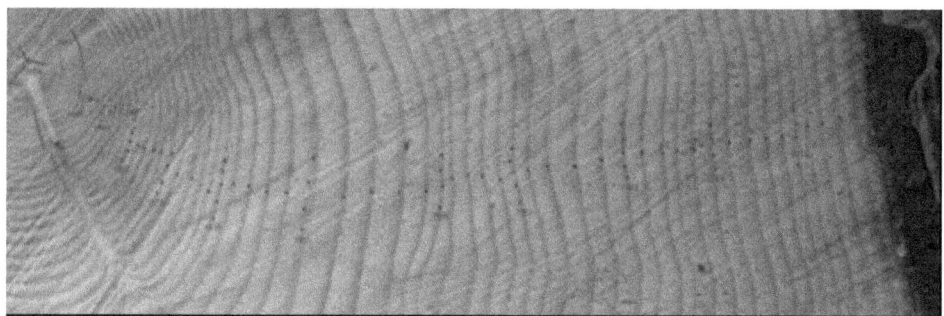

#14 – 2013 to 1979 wide rings, 1952-1979 medium rings, 1952-1938 narrow rings.

In this last group of felled trees near the bike path, there were twelve more trees that I photo'd and with which I counted the rings. The ones shown are in a random order, as I found the stumps walking from one to the nearest next one.

I could find no obvious comparison between the stumps. In general, the rings were wider when the trees were younger and narrower in the last years. Since we don't know when these trees were cut down, we can't determine a year for the beginning of narrower rings.

Conclusion:
I'm really disappointed. After "believing" in dendrochronology for decades, I feel I've been duped. In this last group of stumps studied by the bike trail in South Lake Tahoe, there was no obvious correlation between the rings for particular years of growth. Sure, as a local US Forest Service forester told me, there are factors related to competition that can make

rings in a particular tree for a particular year wider or narrower. A taller tree will get more sunshine to grow. A tree with deeper roots might tap more water than another tree. Mistletoe might have dwarfed one tree and not an adjacent tree.

But after I studied this last group of trees, I began to look for all stumps between South Lake Tahoe and the Emerald Bay overlook, some thirteen miles away. I did see a patter emerging in the bigger stumps. For the first years of their lives, the rings were all wide and all of a sudden the tree rings were narrow for the rest of their lives.

In the Lucky Baldwin tree, which had a known date of death, I was able to count back to the change in width for the rings and it was 1808. Unfortunately (or fortunately), the only people in the area then were the Washoe Indians and they didn't do a lot of publishing of events for particular years.

There was also an old stump at the top of the Emerald Bay overlook and a local forester told me it was cut down "about twenty years ago". That would be approximately 1994. The change in ring width didn't seem to line up going back to 1808 or within five years of that date, though. Then I remembered that they tree was not cut down when alive but when it was dead and they wanted to make it safer for the cars parked in the new parking lot there to not have a dead tree fall on them. So, then it was quite reasonable that the change in rings for that tree could be 1808.

But then I asked, "What event could be so big that all the older trees had the same experience that made such a profound effect that they all started to have obviously narrower rings and continue with the narrower rings for the rest of their lives?" Again, there was nobody but Washoe Indians in 1808.

Then a neighbor, Tim Sizemore (whose cypress tree that was too close to his porch was cut down and put in this), told me of a book that told how the local Indians would purposefully burn down the whole Tahoe basin area to get lush new growth, using the Nevada side of the Sierras as their living source until the Tahoe side regrew. That would explain it. Maybe in 1808 the local Indians burned the Tahoe basin. As the local foresters told me, after a fire, the forest "grew back like weeds" with trees much closer to one another and thus fighting it out for survival, and thus more likely to have narrow rings.

Though, as I said, I was a bit displeased that I couldn't use dendrochronology to date trees in my neighborhood, I realized that there was something to be pleased about. Though the Hohenheim study tried to show that there was 12,000 years of tree rings, that didn't seem to be true. As a bible-believing Christian, I didn't like that the world was older than the 6,000 years that that study supposedly showed. So, in the end, there was no real proof of trees showing more than 6,000 years.

But isn't the title of this book, "Mystery of Tree Rings"? Didn't I just show that there seemed to be a pattern of wide rings before 1808 and narrower afterward being consistent in the area of South Lake Tahoe that I studied?

Yes, I said that. But I point out that in all those trees, there was a consistent display of wide rings for sometimes over a hundred years and then a consistent display of narrow rings for about 200 years. In the Hohenheim study (shown again below), I see now pattern. There is not any extended period of wider or narrower rings that match up. It's a matching of hodgepodge in my opinion, which just isn't up to snuff.

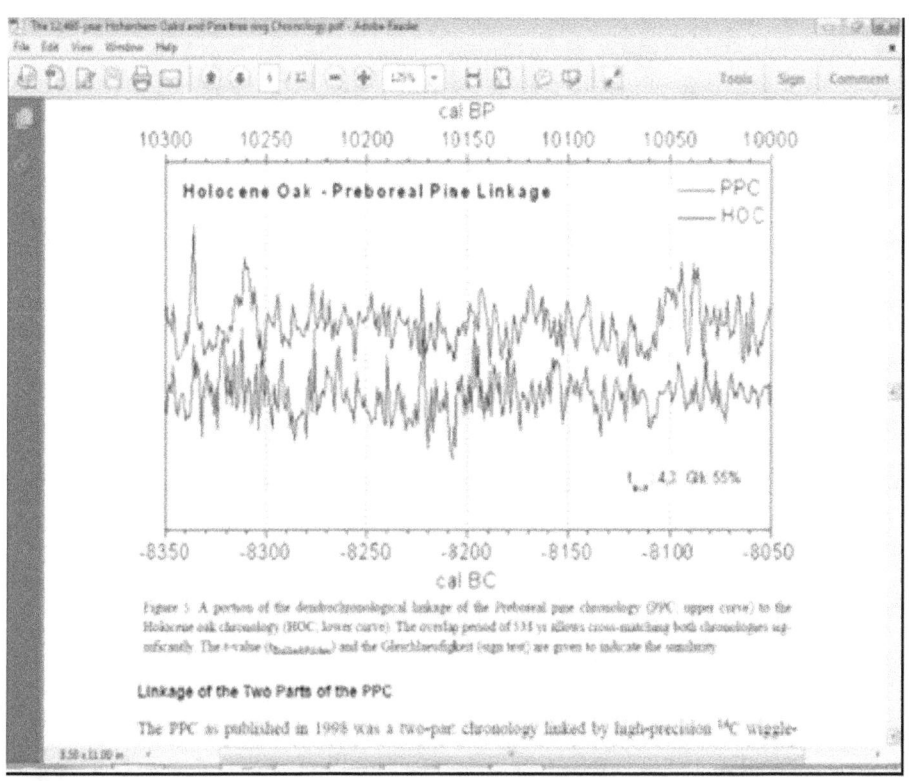